THE SCIENCE OF ANIMAL MOVEMENT

How Birds Fly

BY EMMA HUDDLESTON

CONTENT CONSULTANT
DAVID HU, PHD
PROFESSOR
MECHANICAL ENGINEERING
GEORGIA TECH

Kids Core

An Imprint of Abdo Publishing
abdobooks.com

abdobooks.com

Published by Abdo Publishing, a division of ABDO, PO Box 398166, Minneapolis, Minnesota 55439. Copyright © 2021 by Abdo Consulting Group, Inc. International copyrights reserved in all countries. No part of this book may be reproduced in any form without written permission from the publisher. Kids Core™ is a trademark and logo of Abdo Publishing.

Printed in the United States of America, North Mankato, Minnesota
022020
092020

Cover Photo: Dennis Jacobsen/Shutterstock Images
Interior Photos: Phoo Chan/Shutterstock Images, 4–5; Targn Pleiades/Shutterstock Images, 7; Shutterstock Images, 9, 17, 18, 20, 29 (top); Ken Canning/iStockphoto, 10; Bachkova Natalia/Shutterstock Images, 12–13; Steve Byland/Shutterstock Images, 15; Red Line Editorial, 16; Natalia Paklina/Shutterstock Images, 19; Abdul Rauf Khan/Shutterstock Images, 22–23; Nick Vorobey/Shutterstock Images, 25; Ondrej Prosicky/Shutterstock Images, 26; Don Mammoser/Shutterstock Images, 28–29; DMS Foto/Shutterstock Images, 29 (bottom)

Editor: Marie Pearson
Series Designer: Ryan Gale

Library of Congress Control Number: 2019954242

Publisher's Cataloging-in-Publication Data

Names: Huddleston, Emma, author.
Title: How birds fly / by Emma Huddleston
Description: Minneapolis, Minnesota : Abdo Publishing, 2021 | Series: The science of animal movement | Includes online resources and index.
Identifiers: ISBN 9781532192920 (lib. bdg.) | ISBN 9781644944318 (pbk.) | ISBN 9781098210823 (ebook)
Subjects: LCSH: Children's questions and answers--Juvenile literature. | Birds--Flight--Juvenile literature. | Science--Examinations, questions, etc--Juvenile literature. | Habits and behavior--Juvenile literature.
Classification: DDC 500--dc23

CONTENTS

CHAPTER 1
Moving through the Air 4

CHAPTER 2
Pushing against Gravity 12

CHAPTER 3
Slicing the Air 22

Movement Diagram 28
Glossary 30
Online Resources 31
Learn More 31
Index 32
About the Author 32

Red-tailed hawks live in North America.

CHAPTER 1

Moving through the Air

A red-tailed hawk sits on a tree branch. With a few powerful beats of its wings, the hawk takes flight. It keeps flapping its wings to fly through the air. Then it spots a mouse on the ground.

The hawk flies in a circle, dips one of its wings, and folds both wings slightly back. Its tail is fanned out for steering. The hawk dives down to snatch up its prey.

Wing Shape and Body Size

A bird's wing shape and body size tell about its lifestyle. Hawks and other birds of prey often have wings that come to a point at the tips. The slots between feathers help them glide. The feathers catch rising warm air. This keeps the bird in the air. The birds can also twist and move between trees to hunt or to escape being hunted.

Birds of prey such as kites can make sharp turns.

Small birds such as sparrows tend to have pointed wings that are as big as their bodies. Their wings are strong and quick. They can flit from branch to branch and avoid predators. Other birds with wings much longer than their bodies glide through the air. They don't beat their wings often or fly at high speeds. Flapping such large wings takes a lot of energy. Gliding is easier.

Flying Smart

Birds can sense when the air temperature or wind changes. They stay safe by flying low when rain or storms are coming. When the weather is clear, they fly higher.

Although the kori bustard can fly, it spends most of its time on the ground.

Scientists think there are about 18,000 species of birds. The heaviest flying bird is the kori bustard. It lives in southern Africa and can weigh up to 42 pounds (19 kg).

Hummingbirds are some of the few birds that can hover in place.

The smallest flying bird is the bee hummingbird. It is only 2 inches (5 cm) long. It weighs 0.07 ounces (1.95 g). That is about as heavy as two large paperclips. Science explains how birds big and small can fly through the skies.

Explore Online

Visit the website below. Did you learn any new information about red-tailed hawks that wasn't in Chapter One?

Red-Tailed Hawk

abdocorelibrary.com/how-birds-fly

Lightweight bodies make it easier for birds to fly.

CHAPTER **2**

Pushing against Gravity

There are many *forces* at work when a bird flies. Gravity is a force that pulls objects down to Earth. It takes more energy to lift heavy objects against gravity than light ones. Birds' bodies are lightweight.

They have thin bones and fewer **organs** than some other land animals.

Being light is not the only reason birds can fly. They need other forces to keep them in the sky. The shape of their bodies helps them use those forces.

Takeoff

Birds often use their surroundings to start flying. They may drop from nests or jump from tree branches. Birds that live in water habitats often kick their legs to push off from the water into the air. For example, the noisy, splashing takeoff of swans shows them working hard to get into the air.

Birds create thrust by flapping their wings.

Into the Air

Thrust is one of the forces that help birds fly. Thrust causes an object to travel forward. Birds use their muscles to flap their wings. This creates thrust.

Wings and Lift

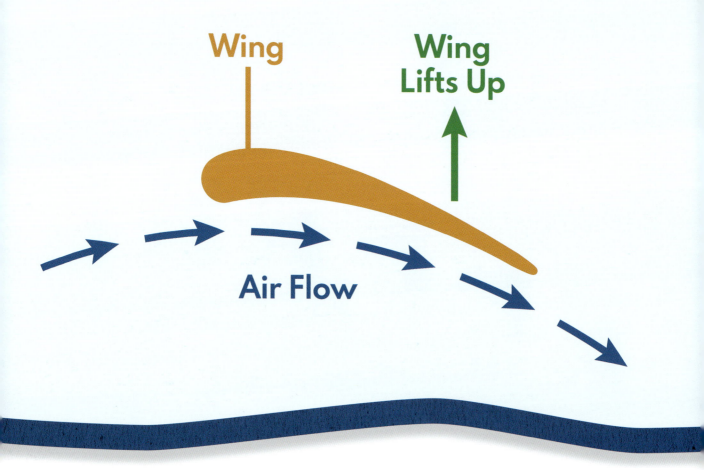

The shape of a bird's wing pushes air flowing under the wing downward. This causes the wing to move upward, in the opposite direction of the air.

To stay in the air, birds need another force. This force is called lift. It is caused by air flowing past a bird's wings. Air may look empty. But it is filled with **particles**, or tiny pieces of things

Like a parachute, a bird's wings catch air.

usually too small to be seen. Those particles help a bird fly.

A bird's wings angle downward in the back. The bird's forward movement forces air flowing under the wing downward.

When a bird flaps downward, it pushes more air than if it held its wings still.

The air moving under the wing creates a force that pushes the bird upward. Flapping pushes even more air under the wings.

Birds with long wings can create a lot of lift. Their wings push a lot of air downward. The wandering albatross has the longest wingspan. Its wingspan is up to 11.8 feet (3.6 m) wide. The bird can travel thousands of miles by **soaring**.

The wandering albatross is a seabird. It soars over the open ocean.

Vultures soar for long periods of time as they look for food.

Soaring is when a bird flies without flapping its wings. The warm air **currents** under its wings support it. Soaring is helpful for large birds because a heavy bird uses lots of energy to fly.

Primary Source

Scientist Michael Habib explains how body size and weight determine how a bird flies:

> Large flyers typically can't [keep] flapping for long periods. . . . They are usually soaring animals, which means that they use . . . other energy sources to stay aloft for long periods of time.

Source: Ella Davies. "The Biggest Beast That Ever Flew." *BBC Earth*, 9 May 2016, bbc.com. Accessed 27 Nov. 2019.

What's the Big Idea?

Read this quote carefully. What is its main idea? Explain how the main idea is supported by details.

The fastest bird recorded is the peregrine falcon. It can dive at speeds of more than 200 miles per hour (320 km/h).

CHAPTER 3

Slicing the Air

When birds fly, forces of resistance such as drag act against them. Drag is when air particles run into a bird. The bird pushes against the particles, which slows the bird down. The bird must create more thrust to keep moving forward.

The faster a bird moves, the more drag it experiences. Another source of drag is from strong winds. A bird uses extra energy to fly through them.

Shaped for the Air

Most birds are shaped so that drag affects them as little as possible. A bird's body shape is streamlined. That means it is shaped to move

Feather Sensors

Birds have different types of feathers. Filoplumes act as sensors. They work similarly to whiskers or antennae. They help birds sense wind and air temperature. They are mixed in with other feathers on a bird's back.

A streamlined body helps birds fly at high speeds.

easily through the air. A pointed beak, round head, and smooth body create a path for air to flow smoothly around the bird.

In order to land, birds fan out their feathers to catch as much air as possible.

Sometimes a bird moves certain feathers to change how air flows over its wings and body. For example, many birds use their tail feathers to steer. They twist their tails to create drag on one side or the other. To brake or land, birds fan their feathers wide against the airflow.

A bird's wings are rounded and thicker in the front. They **taper**, becoming thinner toward the back edge. This helps air flow smoothly around the bird's body. From feathers to bones, birds are made to fly. They move their bodies skillfully through the air, flying from place to place.

Further Evidence

Look at the website below. Does it give any new evidence to support Chapter Three?

How Do Birds Fly?

abdocorelibrary.com/how-birds-fly

Movement Diagram

Albatross

Short, rounded tail for steering

Long wings for soaring

Tapered, smooth shape for ease of flying

Wings slender at tips and no slots

Swallow

Short wings with slots to create power for quick takeoff or hunting

Tail that fans out to steer or slow down

Hawk

Glossary

currents
flows of air moving quickly in one direction

forces
actions that can start, change, or stop an object's motion

organs
body parts that have specific purposes

particles
tiny pieces of something

soaring
a type of flight that requires little flapping

taper
to become thinner or more pointed

Online Resources

To learn more about how birds fly, visit our free resource websites below.

Visit **abdocorelibrary.com** or scan this QR code for free Common Core resources for teachers and students, including vetted activities, multimedia, and booklinks, for deeper subject comprehension.

Visit **abdobooklinks.com** or scan this QR code for free additional online weblinks for further learning. These links are routinely monitored and updated to provide the most current information available.

Learn More

Kuskowski, Alex. *Super Simple Aircraft Projects*. Abdo Publishing, 2016.

Murray, Julie. *Birds*. Abdo Publishing, 2019.

Index

air currents, 20

bee hummingbirds, 11

drag, 23–24, 26

feathers, 6, 24, 26–27
filoplumes, 24
flapping, 5, 8, 15, 18, 20, 21

gravity, 13

Habib, Michael, 21

kori bustards, 9

lift, 13, 16–18

red-tailed hawks, 5–6, 11

soaring, 18–20, 21
sparrows, 8
streamlined body, 24

thrust, 15, 23

wandering albatrosses, 18

About the Author

Emma Huddleston lives in the Twin Cities with her husband. She enjoys reading, writing, and swing dancing. She thinks how birds fly is one of nature's most interesting feats!